Bygones presented by Dick Joice
on Anglia Television

LADY MARGARET BARRY & MRS GERTRUDE SMITH

The Boydell Press · Ipswich

First published by The Boydell Press Ltd
PO Box 24 Ipswich IP1 1JJ
Reprinted 1976

ISBN 0 85115 078 0

Printed in Great Britain by
Lowe & Brydone Printers Limited, Thetford, Norfolk

FOREWORD

My wife, who is a member of the local Women's Institute, had heard from time to time about a very popular W.I. entertainment with the intriguing title *Above and Below the Salt* – an unusual dialogue given by two members, who came from very different backgrounds.

It didn't take much engineering to obtain an invitation to a small village hall to hear what it was all about. The more I heard, the more I realized that this was a 'must', not only for television, but for posterity. After the 'show', I asked the performers, Lady Margaret Barry and Mrs Gertrude Smith to give me a brief history of themselves. This is what I learnt from Lady Margaret as their spokeswoman:

"My friend, Mrs Gertrude Smith, was born in a small terraced house in Norwich in November 1901, and was the youngest of 14 children. She was educated at the Horns Lane Primary, Junior, and finally Senior School. When she left school at fourteen, her parents insisted on her joining her sisters in domestic service, although she herself would have preferred factory work. She remained in domestic service until her marriage in 1933.

Mr and Mrs Smith bought a small cottage in a village near Norwich, where they had neither electricity, nor piped water, and there, their only child, Michael, was born. During the war, Mrs Smith worked on the land, but afterwards returned to daily housework, which she still does.

I, Margaret Barry, was born at Longford Castle, Salisbury in June 1903. I was the seventh child of the Earl of Radnor. I was educated by a governess until the First World War, and thereafter as a boarder at the Godolphin School, Salisbury.

I was married at 19 to a serving soldier in the Coldstream Guards. My husband resigned his commission in 1925, and we emigrated to Southern Rhodesia in 1926, with our 18 month old daughter. We acquired 2,000 acres of underdeveloped land about 40 miles from Bulawayo. Here we gradually built our home and farm through hard times and hard work. Our family grew to six, four girls and finally twins, a pigeon pair in 1938.

We returned to live in Norfolk in 1953, following the death of my father-in-law, Mr W. J. Barry of Great Witchingham Hall.

On joining the Great Witchingham, Weston and District Women's

Institute, I met Mrs Smith, and we served together on the committee. Mrs Smith was a leading light of the Drama Club, so when the 'Speakers' Competition' was organised by the County Federation of Women's Institutes, we entered a team with Mrs Smith as the speaker. I was taking our team to North Walsham to compete in the first round of the competition when Mrs Smith suddenly announced her intention of changing the subject of her talk. I remember saying 'what, at this late hour?', and as we had time in hand I drew into a lay-by, and we planned her new eight minute talk. The subject chosen was 'My first job'. This proved such a success that she went through to the final round and won first prize for our Institute.

It was this talk that gave me the clue to the subsequent success of our joint effort, *Above and Below the Salt*. I remember suggesting it, as well as the title, to Mrs Smith on our way back from North Walsham all those years ago. We have driven miles together, and visited many Institutes, Clubs, Group meetings and so on, and into the bargain made many friends. I could not have had a more delightful and interesting companion than my partner 'Below the Salt'."

The Anglia Bygones film *Above and Below the Salt* was so well received that it's a great pleasure to present it in book form.

Dick Joice

DICK JOICE

LADY MARGARET:
Although there are so many differences in our childhood story, there are things that are very similar too. One is that we both had very good parents and the other was that we were both very happy children, and after all, those are the ingredients that make for a good and happy life afterwards. And I think that is very, very important. I think you may wonder about my parents, if they ever had to do any work. They didn't, of course, have the hard work that Mrs Smith's parents had to do, but they were both very busy people. Amongst other things, my mother started the district nursing service in Wiltshire, where we lived. And of course, that did away with the Sarah Gamps that looked after women when they had their babies; and so Mrs Smith's mother was probably about the last of the generation that had to have help just by the old Gamp from the village.

Another thing my mother did, she was the first president of the Women's Institute in Wiltshire. And I can remember her saying that she thought it was going to be such a wonderful organisation, that it would bring women of all kinds and conditions together, that they would share their problems and their pleasures together, and, of course, we do just that. Now, if it hadn't have been for the Institute, Mrs Smith and I would never have met, and would not have been able to compare all these happy childhood memories!

Now, I was born in a castle in the south of England and I was my mother's seventh child and she went on to have ten of us. Five boys and five girls. Nice tidy kind of family. We were all born in the castle and my father was a peer of the realm when I was born, so even when I was a little child I was given the title which I still have, that of Lady in front of my name. Now If I had married a duke or a marquess, an earl or a viscount, I would have taken his title, but I married a captain in the Army and he is now a colonel, and so I have been able to hang on to my name, and I'm really quite pleased about that. I suppose I've got used to it and it's rather nice to have the same name still. I was, of course very proud when I was known as Norfolk's Lady Margaret in those years when I was County chairman because I don't think there was anybody else in Norfolk would have been known as Lady M, and I know I was called that by most of the people at headquarters and around the County.

When we were children in that great big house, of course, we lived our life in the nursery, but I must tell you first that when I was born there was no telephone in the house and I am told that when my mother was expecting her babies, the stable used to be informed and a pony and cart were kept fully harnessed in the stable ready for word to come from the house that the doctor was needed, and the carriage used to go off and fetch the doctor the three miles from the town. Also, I was told by my old nurse that she was on holiday when the new baby arrived and she didn't know anything about the arrival until she was met at the station by one of the grooms in the carriage, and it was from him that she learned that her ladyship had had a beautiful baby girl. Well, I can't believe I was beautiful. I don't think many babies of a few days old are that, but I suppose that was the word that was put out, so that they should know that I was sound in wind and limb.

MRS SMITH:
There was no horse in the stable waiting for my arrival. We neither had horse, even yet stable. My father was the one who had to fetch the mid-wife. Well, she wasn't really a mid-wife, she wasn't even qualified for that kind of work. She was just a woman who used to come and *do* for women when they had babies. I suppose if she saw complications arising she would have eventually sent for the doctor. Of course, he was the last person you sent for. You just couldn't afford a doctor in those days. And as my father hurried along on this dreadful night in November, the night on the coast that a life boat was sunk and all the men lost their lives, he fell into a hole in the road, (yes, they had holes in those days!) and this hole was three parts full of water, so he got thoroughly drenched up to his waist, but he scrambled out and hurried on and got the woman and brought her home, and he was met at the door by one of my older sisters who had come home to look after the rest of the brood while mother was in bed. And she said, "We didn't want any more babies about here, we've got enough babies as it is". And my father said, "Well, to feel as I do tonight, I think we'll name this one last". And so I became the youngest of a large family, there were ten of us girls, and four boys.

LADY MARGARET:
Now, when you're very small, in a great big house like ours was with all those rooms, oh – over forty-seven – in those days, the children, the little ones, those of six and seven and downwards, would live their life in the nursery in charge of a nanny and the nurse maids under her. Our nursery was a most beautiful big round room in one of the towers

'My father was a peer of the realm.'

with one big window overlooking the front door and another over the back door, and then two looking up the park. Next to the day nursery was the night nursery where Nanny slept and I remember there being two barred cots, one on either side of Nanny's bed, and then there was a third bed where the third child slept, and when number four came along, number three was turned out and you had to move into a room near by; but, in spite of there being I don't know how many bedrooms in that great big house, we didn't have a room to ourselves until we grew up, and I shared with a sister for many, many years. We used to live our life in the nursery; we had our bath in front of the nursery fire. Do you know I don't remember only but one bath in the whole house, and that was in my father's bathroom and nobody else used it. Everybody else, my mother, all the visitors, all of us children and I suppose the servants too, had to have hip baths in the bedrooms, and those not only had to be filled with warm water, but they had to be emptied too. Must have been a most terrible chore.

MRS SMITH:
We had three bedrooms in our house; my parents had one and my sisters and I shared the other and my brothers the other room. If there wasn't room for us to sleep this way up the bed, we slept that way, you see; if there was an extra influx in the family, but we were quite happy. There was no bathroom in our house; I'm quite certain *my* father hadn't a bathroom of his own. In fact, I don't ever remember him having a bath, but I suppose he did sometime or other, and the only time I remember anything about a bath, we always had it in mother's zinc bath which she used to do her washing in, with the two handles, and this particular afternoon, one Saturday afternoon, my two youngest brothers decided to give us two youngest a bath. They thought it would be such a help for mother when she came home and found us bathed and ready for bed. They got the water in front of the fireplace, and they undressed us, but I think we must have stood on the side of the bath because over it went and all the water went all over the floor, and we were then worrying how we could mop all this water up when in walked mother. Oh, she was cross, and needless to say my brothers didn't ever attempt to give us a bath again!

LADY MARGARET:
We used to have to go along to my mother's bedroom every morning of our lives to say our prayers; she must have had a continuous succession of children from seven o'clock, or seven-thirty onwards. And then, after we'd been to Mother's room, we went next door to my father's dressing-room, where he was already getting up because he was always downstairs by eight o'clock in his study, where he did an hour's work with his secretary before his own breakfast at nine o'clock. He was generally shaving, I seem to remember, and I can even hear the noise it made as he stropped his razor: plip, plop, plip, plop, it went and it was a great wide leather strop that hung on the side of the shutter by the window. Then there was the fascination of his money on the dressing table that he'd taken out of his pocket the night before. It was gold and silver in those days because notes didn't come in till the first World War, and there, lying beside his gold watch, was his little silver sovereign case, and we used to fiddle with this, pushing the gold sovereigns under the spring on one side and then you turned this little watch thing over, and then you could put the half sovereign in the other. I don't think

'I became the youngest of a large family.'

we stayed in his room very long 'cause I can remember him saying "Cut along now", and we used to race along the long passage back to the nursery, where we had our breakfast.

Now, I'm going to take you back with my mother to her sittingroom. It was one of the loveliest rooms in the house. It was known as Mother's Boudoir; it was a round room, in one of the towers, with lovely gold and white panelling, and we used to sit down with her at her sofa and a little table and do our lessons. Now, I think it was very unusual for any lady to teach her own children, but she taught every one of her ten children to read and write and I think she was a very good teacher, because I can see even now the print of the books we learnt from. The little reading books and also the print of her beautifully written copies that she did herself for us, and we used to write under them – pot hooks and hangers and little sentences and so on – and until we had read through a little history book, I think it was "Little Arthur's History of England", we were not promoted to the school room.

MRS SMITH:

My parents, when they were first married, they lived in what they called the Mill House, Old Caton. And my father was groom gardener, and he worked for a man who used to buy a tremendous amount of poultry and rabbits and game and all that kind of thing. He was a dealer in that kind of thing, and he used to buy a tremendous amount of geese from Holland and they used to come in at the Old Victoria Station near the Norfolk and Norwich Hospital, and I suppose they used to cut one of the wings because they used to drive these geese through the city of Norwich, and up to Caton. Can you imagine all those geese being driven through just like they did the cattle a few years ago. And then they were put on to meadows and fattened up for the Christmas trade and at Christmas the women of the men who worked for this man, they used to go into a huge barn, the wives and mothers and aunts and daughters and anybody who could pluck, went in and plucked these geese and they used to get threepence a geese (threepence a goose, I ought to say). They were allowed to have the tongues out of all these geese that they had plucked, and they used to take them home and make goose tongue pie, which was a great delicacy in those days, and an extra dish for Christmas.

Now when my father and mother lived at Old Caton they had three little daughters in quick succession and then a little son and then more daughters followed, and when this little boy was about four years old, he caught a severe chill and died, of double pneumonia. Now this sorrow was so great, my parents didn't seem to be able to get over this great shock, and the doctor had to be called in, as they thought they were heading for a nervous breakdown, and the doctor advised my father to change his job and change the house, to lessen the memory of this little boy. And when my father was there he neither drank nor smoked but he got a job in Norwich as drayman for a well-known brewery in Kings Street and he also got a six-roomed terrace house where I was born. And while my father was there he got to like the taste and the smell of beer and I sometimes wonder how many gallons and gallons of beer he must have drank in the forty-eight years he worked for this firm. I know at night he used to come home and as he came round the corner of our house, he'd give a little cough just to clear his throat, you see, so he didn't speak in a muffled way to mother, and she would be sitting there in the living-room, either knitting or sewing for us children, and she'd look up and see his red face and his sparkling eyes and she'd say "You've had a lot of that old sloss again today, haven't you?" but he'd smile and say "Well, no more than usual, you know" and then he'd wash and get ready for his evening meal. And one night I was downstairs alone, I was about six I should think. My mother must have been upstairs, because she didn't ever leave us at night, and I heard my father come round the corner, I heard him

The Countess of Radnor.

give that little cough and I met him at the back door, and, with as much dignity as I could muster, I said "You've had a lot of that old sloss again today, haven't you?", but instead of the soft answer he used to give mother, he said "Don't you dare speak to me like that again, not one of those other girls have ever spoken to me in such a way". I think I rather sobered him up. "Now", he said, "get ready for bed, I don't want to hear another word from you tonight." So a very sad little girl went to bed that night, partly because I hated my father being annoyed at me and partly because I missed the pleasure I used to have every night of counting his money. The money he used to take for the beer, it was all in golden sovereigns and half sovereigns, sometimes they'd be a hundred and sometimes a hundred and fifty pounds. I used to put them all in tens and oh how I loved handling that gold, it really was wonderful, and I think the people must have been more honest than they are today for him to come home with all that money on him, and all the forty-eight years he worked with his firm he was never one farthing out, he always had the proper money, to pay the brewery for the beer.

We were not very smart, we were always clean and neat, but I usually had to wear my sisters' left-offs. They grew out of their clothes so I had to take theirs on and I don't remember ever having anything new, but the one dress that I do remember was a bridesmaid's dress my mother had given to her, and she said it was satin faced cloth, silver grey. Now, I was very pale in those days and very, very fair so, of course, that wasn't really the best of colours for me to wear, but she made this school frock and she put silk collars and cuffs on and gold coloured buttons down the front, and then I ran into school that morning, into the school playground, I thought I was the smartest girl in – well, in the whole school! I never looked so smart in all my life. But one of my friends, well, I don't really think she was a friend, now I come to think of it, because she ran up to me and said "Ha, I suppose you think you look smart?" I said "Yes, I feel smart." "Oh, oh", she said, "you can see it's made out of something else." So I didn't think quite so much of that frock. But, when I come to think of it, I must have had a wonderful mother to have brought us all up as she did and all the hard work she had to do; there was no labour-saving devices in those days.

LADY MARGARET:

We used to have all our meals served in the nursery and that was one of the nursery-maid's many jobs. She had to fetch the meals from the kitchen, which was a very long way off. She had to either carry them up on a tray or else pull them up in a lift, and the lift wasn't a nice passenger lift, or even a small little lift, it was a great, big heavy lift which was meant to bring the luggage up, so it must have been a hard job for her. You may wonder whether the food wasn't always cold, as it always had to come from a long way off. Well, I don't remember it being cold, but I do remember when it was Irish stew we none of us liked it because it stuck to the roof of our mouths, so it must have been cold, 'cause mutton fat is so horrid when it's cold.

MRS SMITH:

We were always well fed. My mother was an excellent cook and she would make the most delicious dinners out of practically anything. She would oh perhaps give us more starch that we should have had, like dumplings and suet puddings and all that kind of thing. And she always used to have a sweet which, oh, I hated the sight of, it was like

'He got a job in Norwich as a drayman.'

a huge snowball, it came on the table, boiled rice in a cloth and it was served with golden syrup, but if she had her eye on the golden syrup, you could only take just a little; well, that wasn't so good but she said it was nourishing. I could have had school dinners and I, being one of a large family, could have had them free, but I wouldn't allow my pride to go for school dinners, I'd rather come home, and not only that I heard that they gave boiled rice some days, and I didn't like that.

LADY MARGARET:

After Mother's lessons, almost every day of our lives the nursery party would go out for a drive in the wagonette. We used to look upon this as a terrible chore; I'm almost ashamed to think that we thought it was a chore now. But the carriage would come round to the front door; it was a wagonette, open wagonette, and on the box was Fred Martin, the coachman, and beside him sat Leonard the boy. Both of them wearing full livery, top hats with cockades at the side with this lovely coat, white breeches and top boots, and the moment the carriage drew up at the door, Leonard would jump off the box and run round to the back, open the door and there stand, strictly to attention, and let us all clamber in, and then poor Leonard had to stand and hold the baby while Nanny got in, and hand the baby up to Nanny. I can see Leonard so clearly in my mind; I suppose he was about fifteen or sixteen; he wasn't very big and the coat he had to wear, this beautiful livery coat, was rather too big for him and he had a pale spotty face, but he was such a nice boy and I often wonder – I'm afraid he must have been killed in the First War because I never remember seeing him afterwards, and most of them went to the War, you know, and so many of them never came back. Then we used to go through the park with Fred's horses, as we called them. Fred was the coachman, trip, trot, trip, trot . . . up on the downs; it was good for the children to get out of the valley which was supposed to be rather damp. Oh how bored we used to be! We played endless games of 'cats cradle' and 'I spy' and all the rest, but, oh how glad and thankful I was when I was old enough to go riding instead of driving. I always loved the riding. To start with, of course, we were led along by a man with a leading rein and then we were promoted to riding with a groom on horse-back. I never had to fetch the pony, feed it, I didn't know how to bridle or saddle it. I just waited at the door, all beautifully togged up in my lovely habit, cap and whip and all, and then the pony would be brought round to the door and I was lifted on to the pony and through the park we went and we then came to the gate and in a lordly voice, I'd call out "Gate, Mrs Horting", and Mrs Horting came out and dropped me a curtsey and threw open the gates and then a very nice voice, I said "Thank you, Mrs Horting", and through we went, through the village and with everybody pulling their caps and dropping us curtsies, and on to the downs and away we went. When it was Charlie the groom that took us we went like the wind because he was young. I think he was my first love, was Charlie. But it was quite often Mr Baker, the studgroom, and he was a rather a grand gentleman, who came in a top hat and not a bowler hat. He was very portly and we went rather sedately, but even with Mr Baker I loved those rides. Oh, what a lucky little girl I was. I'm sure I couldn't have had a happier childhood.

MRS SMITH:
Although my father had this drinking habit, he really was a good father and if he could take us out to shows or the fair or circuses or exhibitions he was more than pleased to be able to, and he would also take us down to the sea one day a year, and I wasn't bored with *my* carriage ride because we only had one; he'd let us have one ride along the whole sea front and back again. We used to feel like royalty going along, and the tears there used to be when we had to get out. He said he can't only afford one ride, and but oh how we did enjoy that.

'I wouldn't allow my pride to go for school dinners.'

I suppose I've been into nearly every public house in Norfolk. The old ones, as I used to go with my father, nearly every day during our summer holidays and he would stop at practically every public house we came to, even if he hadn't beer to leave he used to say he had business to attend to; I don't know what business it was, unless it was to try their beer. It might have been, but when we came home at night we'd both be very tired and we'd both be asleep. The reins used to fall on the horses' backs and they'd find their way back to Norwich, and then they'd turn in at the stable gate, never once hitting the bars or the gate posts, and we'd wake up with a start when their hooves hit the weigh bridge and the wheels went over; we used to wake up and I'd jump off the waggon and run home to tell mother that father was on his way and he'd water the horses and feed them and bed them down for the night. But, how I did enjoy those rides. He used to tell me all sorts of stories, and we used to be very, very happy.

LADY MARGARET:
I think, as children, we probably saw more of my parents than many children of the same sort, living under the same sort of circumstances. I think they were very family conscious. We always used to spend the evenings with them. In the winter we went down to the drawing-room and played games, or if there were people in the house, we had to play quietly by ourselves. In the summer we went into the garden with

them, and went into the kitchen garden and were given fruit and all the rest. We never went downstairs in our day clothes, we had to be togged up in the most beautiful little white dresses and ribbons and the passages were so cold that we were wrapped round in shawls to go downstairs.

I must tell you one more thing about my childhood and that was when my parents were entertaining in the house. I think the household must have enjoyed it just as much as we did. The house was full of people, every room would have been occupied, the household generally was stretched to the limit and putting their best leg forward to make everything appear to go on oiled wheels. I do really think it did. I think the servants were very, very clever and careful about how they ran things; but what we children really enjoyed was being allowed to sit up in the gallery in the dark after we had had our biscuits and milk in the nursery, in our dressing-gowns and see the grown-ups go into dinner. It was up in the dark, this gallery, overlooking the central hall and there was a beautiful big chandelier of lamps in the middle of the hall. First of all Mr Warman, the butler, came through, very pompous man, Mr Warman, we never liked him. I forgot to tell you Mr Warman didn't live in the house, he lived in a cottage, just by the other end of the garden, and he used to come through a little path, I suppose early morning and late at night, I don't know, but he certainly didn't live in the house and we didn't like him. First of all, in he came, through the hallway, he'd throw open the big double doors and then, in the most pompous voice, he'd call out "Dinner is served, m'lord". Then he'd go back through the hall, down to the dining-room. Then through that door, first my father with a lady on his arm, and then the rest of the party, two by two with my mother and her escort right at the back. And you can imagine the excitement of the lady's maids, and the nannies, and people who were standing behind us, straining their necks over the balustrade, and whispering like anything, whose lady looked the finest, and whose jewels were the best, and so on. Oh, we children thought it was wonderful! We never tired of it, it was just like a sort of pageant or play going on; it was all over really in a few minutes, almost seconds, but we never, never tired of that. It was great, it was a thing to look forward to, from one party to the next. Then, of course, we were hurried up the stairs, they were circular stairs, and tucked up in bed and that was that.

MRS SMITH:

We children loved my father coming home when he had had enough to drink because he used to teach us to dance and in the summer we

'The carriage would come round to the front door.'

danced outside and in the winter we danced inside, and the only music we had was a small Jews-harp which he used to play, and he'd pull his thick woollen stockings over his trousers so he'd look a cross between a Volga boatman and a ballet dancer, and then he'd teach us all the steps of the dancing and oh, how we loved it. It really was wonderful to learn dancing with father but of course he always had to have a certain amount of beer, if he hadn't had that amount, well we went off to bed, feeling very sad.

One thing which greatly amused us was when my sister, my eldest sister, she was head-housemaid, used to come home, it was more like a lady coming home. Oh, she used to be so smart and she had such wonderful manners and everything, and she used to bring her green Riley bicycle home. Now, to have a bicycle in the family, for working-class that was really something you could brag about to your school friends, that you had a bicycle in the family. Especially a green Riley. I don't know what she brought it home for, because I didn't ever see her ride it. It used to be pushed into the front room and there it stood against two chairs in front of the fireplace. We could look at it through the window but of course the front room was more or less sacred ground to us children. We were not allowed in there, only just on very special occasions. Now this same sister used to take my mother up to Barnes store, because in those days, if you spent a certain amount of money you got a free glass of port wine, and she used to take my mother up there because it was such a luxury for her to have anything like that, and then she would take her out into the city, just to have a cup of tea for a little treat. Well, of course, that was the signal for us children to get into the front room with that bicycle, and we did everything to a bicycle you possible could, except ride it. We used to push it all round the room, we'd ring the bell, and pump the tyres and turn the pedals, and we'd also have all our friends come in, so they could really see we had a bicycle in the family, and they'd ring the bell and turn the pedals and do everything they could. But they were all cleared out before mother got home and the door shut. I don't think she ever knew we had all those in the front room to see that bicycle. And the same sister used to bring a harp home. Now whether she brought the bicycle home to carry the harp, I never did really know, but this used to be carried upstairs and then it would be brought down each evening and she would sit on a chair and us children would sit round on our stools, and she'd ping it; and we used to listen to that as though we were listening to the 'Warsaw Concerto' . . . oh, how we did enjoy that evening meal, and that evening music! It was nothing we could sing to because it was only just a ping, but we felt it was lovely.

LADY MARGARET:
After breakfast, those of us that were able, I suppose the three, four and five year olds, were allowed to go downstairs to be in the dining room while the grown ups had their breakfast. It was a time when we were to be seen and not heard, and if we made a noise we were turned out. I think we must have been turned out very often. Because I remember being out more than I do being in. But we didn't mind being turned out because the house was so busy at that time of day. The house-maids were finishing off the front rooms which had to be all cleaned and tidied by the time the grown-ups came out of breakfast and then the 'bothy boys', the garden boys, were all in the house and they were doing the flowers; they used to come in with a great big carrying tray on legs and they had this laden with beautiful flowers and potted plants and they did all the big arrangements in the drawing rooms and the sitting rooms, and they used to replenish the stands of potted plants in the hallways and the lobbies. And we loved to help them; can't you imagine what a nuisance we must have been with the watering can and I don't know what. We used to know them by their

'We were led along by a man with a leading rein.'

christian names and they were such nice young men, we never had any trouble with them and they allowed us to run round with them every day. And then there was another person, we liked to help him too, and that was a Mr Goldsack, my father's valet. He used to really look after my father's clothes, to see that the right suit was out on the right day and his clothes were always cleaned and tidy. But he had other jobs, he always served at table and he also did the writing-tables. Now, doesn't it seem funny that anybody had to *do* the writing-tables? But, do you know, there was an awful lot of writing-tables. You see, there was no telephone communication and the people who used to come and stay with us were very leisured people and they used to write letters to all their friends and relations and keep in touch with them in that way. And Mr Goldsack was ever so pleasant to us children, and we liked him dearly. He used to have a great big basket full of writing paper, and pens and inks and paper and all the things that were needed for the tables and of course, we loved to pinch a pencil out of his basket and we thought they were his pencils and never really, were clued up to think they were my parents' pencils.

At ten o'clock my mother would go through to the back to do her housekeeping. Now, we used to first go through a swing door and there, waiting for her outside the steward's room, was the house-keeper. She was a very grand lady, very small and I remember her so well, she was dressed in a very tight fitting grey dress, she had a black bow in her hair and she was very prim and she used to address my mother and they used to have such a dull conversation, we never waited to hear what they had to say. We raced down a few stone steps and went round the corner into the still-room. Now, I don't think there are many still-rooms left today, but in those days they were the most fascinating kitchens where all the baking for the house was done; all the bread and the biscuits and so on and there was always a lovely smell of new baked bread and there was nearly always a bowl of rising dough in front of the stove, and we used to poke this dough, I can remember picking up the green-grey blanket that covered it and poking the dough with my finger. Mother would call us all too soon because she didn't have to go and see the still-room maid; the still-room maid was under the jurisdiction of the house-keeper, so anything that she wanted to have said to her was said to her by the house-keeper. Who, as I told you was a very grand lady. You see, all those top servants in the house were very grand, and it's curious to think back now that I was really brought up and taught my manners by those servants, there was a wonderful sort of friendly feeling and yet a real respect between them and you. We were never allowed to address them angrily or crossly, we had to address the top servants with their title, Mr Warman, the butler; Mrs Quinton, the house-keeper; Mrs Davidson, the cook and Miss Crisp who was my mother's lady's maid. And they were very, very grand, and if we'd been cross or tiresome to them they'd have told Nanny and Mrs Cutt, the head nurse, who would have given us a really good talking to, and we never really liked that too much 'cause Nanny was very dear to us and, if she was cross thus, we really minded terribly. Well, when Mother had had a word with Mrs Quinton and she called us from the still-room, we had to go with her down the back passage, which was a great big wide passage, into the kitchen; but it wasn't just into the kitchen that we went, it was into a whole apartment of kitchens. You went through an archway and there were two larders, one on either side of the door, one a meat larder and the other a game larder, and then, in front of you, was a very big scullery where the scullery maids did all the washing up, and you went round the corner into the kitchen, which was a great big, very lofty room with windows round the top part and the walls just lined with copper pots and then, behind a screen which was called the hot plate, because the plates were put on the far side of it to warm up, there was a great big open fire with a spit at the side. Well now, I never saw the spit being turned, I suppose it was too early at ten o'clock in morning, to start cooking the meat, but that spit was turned by the kitchen boy, and the kitchen

'The most beautiful little white dresses.'

boy that I remember was an old, old man. He was dressed in a striped suit, rather made of the sort of material that a butcher's apron is made of, and he was very stooped with all the work that he had done all his life, carrying the coals and all the rest. He was a very dear old man called Jesse, and when he was retired he lived in a cottage in the park and we used to go to visit him as children. Now, Mrs Davidson, the cook, was there waiting for her ladyship and she was very grand, very large, if I remember, with a white cap on her head, which made her look even taller. And she used to take my mother through, even further on still, into what was called the cook's parlour, where Mother did her house-keeping. And do you know, all she had to do was to go through the menu for the day and then she'd mark on the side of the menu, which was beautifully written out in a sort of copy book, how many people were coming to each meal and whether she approved of the dishes, and occasionally, I think, she used to scratch something out, perhaps his lordship didn't like it or it was too big a meal for the family, or what I don't really know, and then they would discuss any party that was coming on in the near future. We were given a bit of chocolate, a very delicious sort of Meunier chocolate or something, and then back we went along the back passage to my mother's sitting room.

Now, if we'd gone further still down the back passage, we would have passed two funny little rooms that were known as the 'lamphole' and the 'boothole'. They weren't really holes, they were little rooms. The 'lamphole' smelt disgusting, because it was where all the lamps were filled with oil, and I can remember the smell now. And the next door was that of the 'boothole', where all the boots were cleaned, and that didn't smell so bad. But we never really went down the back passage, further than that but if we had, we should have come to the pantry where the men-servants worked. They worked in baize aprons and shirt sleeves, cleaning all the silver. And over the door of the pantry there was a long board with a row of little bells hanging, with the names of the different rooms underneath and when that bell tinkled, then the footman would pull on his livery jacket very fast and race through to the front to see what was needed. Now the footmen were the only ones that wore the livery. It was a very smart livery, it was a dark plum colour jacket, with brass buttons and a striped waistcoat with red and gold, and they had black trousers. And if they were serving at dinner in the evening, they wore knee breeches and black stockings and buckled shoes. My father never allowed the men servants to wear gloves, when they served meals, he said that he'd sooner know that their hands were clean rather than they might be dirty inside white gloves. So he never allowed it. They never had powdered wigs in my home, but I did see a powdered wigged footman, two of them once. I was staying with my grandmother in London. She lived in Cadogan Place where that big hotel is now, and we used to go past Cadogan House which is also no longer there, it used to be where that big block of flats that looks like a gasometer stands. And we went past this house and I can see it as clearly as anything, there were two liveried footmen, they had a bright red, very smart uniform and these white wigs on their heads, and they were standing very strictly to attention. They must have been expecting some very smart visitors. But that never happened in my home.

MRS SMITH:

All my sisters, as they left school, they were found places in domestic service, but I, being the youngest, I thought, at least, they'd make an

'A six-roomed terrace house where I was born.'

exception to the rule. So with the help of my teacher I was able to obtain a job in Harmers Clothing Factory. There I was going to learn tailoring. I was very fond of needlework, and the teacher got me this job. So of course, when my mother began to speak about what she was going to do to find me a job I said "Oh, there's no need to worry, I've already got a job". "Oh" she said "Indeed, where is it?" and so I told her. "Oh no" she said "You're going out to service, like the rest of the girls, we can't treat you any different from the others, you too must go out". And I didn't want to. My sisters all went into good Norfolk families, and if they liked the housework, they went as under-housemaid and worked their way up, and if they liked the kitchen, they started as scullery maid and worked their way up to that. But I didn't want to go into a place like that. I didn't want to go to a place where perhaps my sisters would tell me how to do things and show me how to do them; at fourteen I thought I knew everything. So I decided that, if I had to go out to service, which it now seemed inevitable, I'd go somewhere where there was only me kept. Somewhere where I could be queen of the kitchen. Head of the house, do exactly as I liked, when I liked and how I liked. Of course, the lady of the house never entered into my mind, I thought she'd just be a figurehead, and I would carry the whole household on my shoulders. So I answered an advertisement, it said – 'General servant wanted, must be young and able to cook.' Well, I was young and I was able to cook after a fashion, but what this general servant meant I wasn't quite certain about that. I read in the papers where the generals lead their men into battle, and I thought, well it can't mean that because there'll only be me there, so I shan't have anybody to lead. So I answered the advertisement and I went for an interview and there was a lady there who was a char and she lived in a room in the next road to me, and I thought "A link with home, it won't seem so far away". And she said, "I'll tell the lady you're here". So she went and the lady came out and greeted me, and she said "Follow me into the drawing-room". So I followed her through the hall and I couldn't help noticing her lovely black dress as she walked, and it rustled and I thought "Oh to work for someone with a frock like that, one that rustled". And I thought "When I get old enough and enough money, I'm going to buy myself one of those frocks what rustle". But I've never had the figure not to carry it out, so I've never had a black silk dress. I followed her into the drawing-room and she sat down and she told me to sit, and she asked me all sorts of questions, could I do this and could I do that, and all that, and I said "Oh yes", even if I couldn't. I thought "Well, I don't want to lose the job. I'll try and find out how to do the things even if can't do them". So she said "Now, when can you come?" So I said "Well, I can come on Thursday, Thursday afternoon". So she said "Oh very well, come soon after lunch, and then I'll show you where the tea things are kept and how I like tea served. And then there would be dinner at night, eight o'clock."

Wednesday I spent packing my bag, or rather my dress basket. I expect some of you have met up with a dress basket, haven't you? Where the lid holds just as much as the bottom. And I collected everything I got in the world and I put it in this dress basket; the result was the lid was as full as the bottom. But I strapped it up, and I half carried it and half dragged it downstairs and my mother said "What on earth have you got there?" I said "It's my luggage I'm going to take tomorrow." "Oh", she said. "There's no need for you to take that" she said. "If you'd pick out the things you'd need until about Sunday, the lady is sure to let you out, you may be able to come out for tea." Because it was only up the Thorpe Road you see. "You pick out what you want, and wrap it up in a brown paper parcel." I said "Take a brown paper parcel to a new place? Oh no", I said. "I'm not going to let them think I haven't got anything." "Oh very well then" she said. "You take that basket then." So the next day, off I went to this place and the lady showed me upstairs and she said "You'll find three uniforms in the wardrobe. There's a black one and two silver grey. You can put the black one on for this afternoon, as there won't be much

The Picture Gallery, Longford Castle.

dirty work." And, after I unpacked, I got this black dress out and I put it on and it hung more like a clergyman's cassock. It had been made for someone with much bigger statistics than mine was. But I put a belt around the waist and I dragged it upwards and then I put an apron on and then one of those bonnet shaped caps, I tucked all my hair in and I tucked it all in and then I looked in the glass at the finished article, and I thought "Here, indeed, is a full blown skivvy." I think that's what turned me from domestic work because all my school friends went into factories and offices and shops and they used to say to me "You're not going to be a skivvy like your sisters, are you?" And I used to say "Oh no, I'm going to learn tailoring." But here was I, a skivvy, and I thought I better go downstairs and see where these tea things are. And I remember when I got to the bedroom door and I looked round I thought "Oh, at last, a bedroom of my very own and a bed I wouldn't have to share." And I went downstairs feeling on top of the world. And I got the tea ready and I cooked the dinner, and everything was a success, beginner's luck, must have been. No burnt offerings or anything. And the daughter, who lived with her mother, came out and she said how beautifully I had cooked dinner, and she took the coffee back into the dining-room, and then she came out with the cups and then she said "Now you'll wash them up, won't you, and tidy the kitchen and have your supper and then I should go to bed, you're not very old, and you mustn't stay up very late. I'll wind your clock up and I'll set the alarm for half-past six, and you can get up and set the fire in the dining-room and get your breakfast ready and everything." So off to bed I went, and I got into bed, but that didn't seem so nice then, when I got into bed, I missed the bodies each side of me, you see. Being the smallest and the youngest, I usually found myself in the centre of things, and I missed my brothers in the next room, they always used to tell us riddles and stories and all sorts until we fell asleep. But, anyhow, at last I fell asleep and the next thing I remember was loud knocking on my door and a voice which I recognised as the daughter saying "Aren't you getting up today? It's past eight o'clock." I hadn't heard the alarm, or it hadn't gone off, but I was still sleeping when she knocked. But I scrambled out of bed and I hurriedly dressed, I put the bluey-grey uniform on, you can imagine what I looked like going downstairs, I hadn't time to adjust it, anyway – so you can form an idea what I looked like. Down I went, and I helped as much as I could, to catch up with as much as I should have done, and it was while I was trying to eat my breakfast that I realised how lonely I was, and presently the tears began to fall and I got up to wash up the things that were dirty, and the tears were dropping in the washing up bowl, making circles, you know, like they do when you throw a stone in. And when the lady came out to give me orders for the day I was weeping bitterly. She thought it was because I layed late. I said "No, it isn't that", I said "I don't like being here." I said "I shall have to go home, I shall have to go home today. I can't stay here." "Oh dear", she said, "What a trouble you domestics are, I must go into the city and see if I can get another girl. Will you stay here until I come back?" I said "Oh yes, I'll stay." And she came back about four o'clock and I got all her tea ready and she had her tea and then she came out and she said "You know you cooked the dinner so beautifully last night, will you cook it again tonight?" So I said "Oh yes, I don't mind." And she said "You'll find a pheasant in the larder, have you ever plucked a pheasant?" I said "Oh no I haven't plucked a pheasant, but I have plucked a bird." Which I hadn't but I thought I better say I had, but I thought "I shall learn, no doubt, there's always a first time." But I don't think I should have been so anxious to have offered to pluck that bird if I'd known what a high standard that was. Oh dear, that was dreadful. But I plucked and got it in the oven, and cooked it, and the daughter came out after for the coffee and she said "You'll wash all up before you go won't you?" And I said "Oh yes." "And you'll get

Lady Margaret and two others with net.

the sticks and the coal in for tomorrow?" "Yes?" I said. "And those shoes along the dresser there, they all want cleaning, will you clean them before you go?" and I said "Oh yes." And she said "Mother will be in the dining-room, and will give you your wages." I said "Oh thank you."

During the day, I found time to pack my dress basket and that stood in the hall waiting for the take off, and I went into the dining-room and the lady stood there still in her lovely black silk dress, and she said how sorry she was I was leaving, as she thought I would have made her a splendid little maid. And she said, "I've taken two of your meat coupons, and here are your wages, one and six." I said "Oh thank you very much", and she said "You can go out the front door, it's lighter than the back." And as I was going through the hall, the daughter said "Mother, you're never sending that child home at this time of night, are you? It's past ten o'clock and the last tram car has gone you know, and it's raining hard." But I didn't wait to hear what Mother said, I went out of that door and down that drive as though there was somebody behind me, and I got onto the Thorpe Road and I walked, and I walked, and I walked with this heavy old basket and

I thought "Oh, if only I'd followed Mother's advice and had that old paper parcel." Because I couldn't stand it down anywhere, you see it was so wet I had to keep on holding it, even when I stopped to get a little breath. But by the time I got home it was past eleven. My father was home, he'd been late as usual, and was having his evening meal, and he looked up when I went in and he said "Oh you're back again." I said "Yes, I didn't like being there. I was so unhappy, and she's taken two of my meat coupons." "No" he said, "Never mind about the meat coupons" he said, "This is the place to come if you're unhappy" he said "Never mind, old partner, you'll have to come and be my servant."

So I was at home for a week or two, and I thought my parents had relented and they were going to let me stop at home, but there was a vacancy occurred as the tween-maid where my sister was cook: this vacancy was between the house and the kitchen. My mother said "This would be the very place for you", she said "You will then decide which you'd like to do best, the house or the kitchen work." I didn't want to do either, but I went. And now for this my mother didn't exactly make me a trousseau, but she made me three chemises, of grey calico trimmed with torchon lace all hand done and feather stitched all round the neck and you know what you wear with them . . . I can't exactly say, it might shock the gentlemen. When I got to this place I found to my amazement that the other girls wore combinations. So if that saying had been then, I still wasn't 'with it'. But I wasn't to be beaten, my afternoons out I took these two chemises and what went with them and I made them into combinations. I don't think Norman Hartnell would have called them combinations, I think he would have said they were a combined effort. But I was really pleased I done it. But I kept one lot intact you see, because when I went home for the weekend or anything I didn't want my mother to know that I'd cut them about and made them into combinations, so I used to wear the chemise you see, the one chemise I got left. She said "I'd thought they'd be much whiter by now." But, of course, I didn't never wear them, you see, not that particular pair, only when I went home. So that's why they weren't so white as she thought. Well, now my sister, who I lived with, she was one of our favourite sisters, and all the other girls envied me going to live with her. But a sister who comes home on holiday and sister who has rather a wayward girl to train is quite different and, oh she was sharp. She said "If I can't make you into a good servant, nobody else will." And I'll just tell you what happened to me one day. I went down into the kitchen and I said "Is there anything I can do to help you?" "Oh no" she said "I've done it all. I've got all the breakfast already."

Mrs Smith's parents.

she said. "You can take a tray into the drawing-room and you can bring the silver out, and you can clean it in the pantry." She said "You know, they're spring-cleaning in the drawing-room." And I said "Oh yes." So I went and got them, and amongst the silver things was two little stags, they were on little pedestals, and their legs were like pins. Well, I proceeded to polish them as though I was preparing a stallion for the Norfolk Show. Well, the result was, there was two pieces in my hand, the pedestal and the stag. I went out into the kitchen and told my sister what had happened and she said "Oh dear, I don't know what the old lady will say to that; I shouldn't be surprised if she doesn't send you home." "Oh well" I thought to myself, "that's just what I would like. I don't mind that bit." So she said "You must be careful!" So I finished the silver and then after we'd had breakfast I went upstairs to do the old lady's room. Of course, that was the time of marble-top wash stands and basins and everything. And the housemaid had told me that whenever I went into her bedroom I should always fling the window open and open the bed. Well, I was thinking so much about this stag, I forgot to do either, and I was busy emptying the water, I got the bowl and was emptying it into the pail, when she came into the room. She must have wore rubber soles, because I didn't hear her, and she shouted at me because I hadn't opened the window. Down went the bowl on the pail and it broke in halves, one of the best bowls in the house! And, oh dear, the water went all over the floor and I went and told my sister. And, oh she did go on about it, but at any rate, I finished the bedroom, mopped all the water up, and then I came downstairs, and then I said "Now what do you think I better do?" She said "I should go into the drawing-room and see if they'll give you something to do." So in I went, into the drawing-room and the housemaid said I could clean inside the windows. She said "Get the things to clean with and the steps you'll find so and so", and I went and got them. Now whether I stood those steps in rather an awkward position, but when I go to get up, I must have thought the windows were elastic and would go out, but instead of that *this* part of me went through the window, and it smashed that huge window, and all I could remember was the gardener, who was cleaning the outside, and he looked through the hole I'd made and he said "Now, what have you done, little 'un?" He always called me 'little 'un'! "Oh", I said "I don't know. Isn't it dreadful!" My sister thought it was a bomb had dropped outside on the lawn, she came running in and when she saw that she said "Oh my goodness!" she said "Whatever will the old lady say? You must take her coffee in and tell her all what you've done." "Well", I said "I'll pay for everything."

Eleven o'clock came round very quickly that morning and into the dining-room I went with the coffee, and the old lady was sitting in the armchair and she was well over eighty. Little lady, she was, and she looked up and she said "Oh, Gertrude, what do you mean, bringing my coffee in?" So I said "Well, Madam, I've come on rather a sad mission." I can tell you I wasn't laughing, I assure you. I said "I've had some terrible accidents today." She said "Oh, what are they?" And I proceeded to tell her all what I done. I said "And I'm quite willing to pay for them and I'll have the little stag soldered and I'll pay for the window and I'll pay for the bowl." "Oh dear", she said "How long are you going to stay here, as between-maid?" She said "Don't worry about that" she said, "Although I look an old lady, I was young once and I used to do the same kind of things." She said

'What a wonderful childhood I had.'

"You needn't pay for anything" she said. "It's quite alright", she said. "Accidents will happen, and I'll have the window fixed. Thank you, Gertrude, for bringing my coffee now. Don't worry about anything, it will be quite alright." And I thought "What a dear old lady, I'll do anything for her while I was there." And I was there for just over two years. But I think partly why I stayed was because I had rather a crush on the garden boy.

LADY MARGARET:
What a wonderful childhood I had! Good, kind parents and a wonderful home to live in, surrounded by beautiful things. I'm almost ashamed to say I was born with a silver spoon in my mouth.

MRS SMITH:
We thoroughly enjoyed our childhood, too, although we weren't very well off. I'm sure I wasn't born with a silver spoon in my mouth, but whatever kind of metal it was, it was cleaned every week. That was how I got my pocket money; farthing a week by cleaning all the cutlery and running the errands, and I could do exactly as I liked with that. I could save it, or spend it, but I can assure you that I used to spend it!